Reaching for the Moon

The Apollo Astronauts

Explorers of New Worlds

Reaching for the Moon

The Apollo Astronauts

Hal Marcovitz

Chelsea House Publishers
Philadelphia

Prepared for Chelsea House Publishers by:
OTTN Publishing, Stockton, N.J.

CHELSEA HOUSE PUBLISHERS
Production Manager: Pamela Loos
Art Director: Sara Davis
Director of Photography: Judy L. Hasday
Managing Editor: James D. Gallagher
Senior Production Editor: J. Christopher Higgins
Series Designer: Keith Trego
Cover Design: Forman Group

First Printing
1 3 5 7 9 8 6 4 2

Library of Congress Cataloging-in-Publication Data

Marcovitz, Hal.
    Reaching for the moon: the Apollo astronauts / Hal
    Marcovitz.
        p.   cm. – (Explorers of new worlds)
Includes bibliographical references and index.
ISBN 0-7910-5957-X (hc) – ISBN 0-7910-6167-1 (pbk.)
1. Project Apollo (U.S.)–Juvenile literature.
2. Astronauts–United States–Juvenile literature.
[1. Project Apollo (U.S.). 2. Space flight to the moon.]
I. Title. II. Series.

TL789.8.U6 A5538 2000
629.45'4'0973–dc21

                                                00-034597

Contents

Fire in the Cockpit

The crew of Apollo 1, *Virgil I. "Gus" Grissom, Edward H. White, and Roger B. Chaffee. Shortly after this photo was taken, on January 27, 1967, the three astronauts lost their lives in a tragic fire inside their space capsule.*

Virgil I. "Gus" Grissom, Edward H. White, and Roger B. Chaffee were assigned to a routine job in the late afternoon of January 27, 1967.

As the hazy sun hung over the Florida sky, Grissom, White, and Chaffee ascended to the *Apollo 1* space capsule. Their small capsule was atop a **rocket** that towered more than 360 feet above Pad 34 at Cape Canaveral. The **astronauts**, wearing their heavy flight suits, were strapped into their seats. They were preparing to run through a test of the rocket's ignition. Their mission, the first of the Apollo program, was scheduled for liftoff in three weeks.

As part of the test, the capsule (known as the *command module*) and the massive rocket it rode on were unhooked from the life-support and power systems provided by Pad 34. Inside the command module, the astronauts were on their own.

Problems had plagued the test from the start. The radios that allowed the astronauts to communicate with Mission Control were not working right. The climate control, which regulated the temperature inside the capsule, was going haywire as well.

The mission's commander, Gus Grissom, was especially upset. He had been an astronaut since the earliest days of the space program. Certainly, problems with the complicated equipment used to take men into space were nothing new. But there had been so many problems with the *Apollo 1* capsule that the veteran astronaut's patience was growing thin. One day, Grissom had hung a large lemon inside the capsule to show his disapproval.

White and Chaffee did not share Grissom's pessimism. White had flown once before in space. He had been the first American to walk in space. "You have to understand the feeling that a pilot has, that a test pilot has, that I look forward to the first flight," White said shortly after he was assigned to the crew of *Apollo 1*. "There's a great deal of pride involved in making a first flight."

Chaffee, who had not yet flown in space, was equally enthusiastic. "I think we've got an excellent spacecraft," Chaffee told reporters a few weeks before the January 27 test. "I've lived and slept in it. We know it. We know that spacecraft as well as we know our own homes, you might say. Sure, we've had some developmental problems. You expect them in the first one."

The test dragged on, and the communications problems continued. "How do you expect to get us to the moon if you people can't even hook us up to the ground station?" Grissom complained. "Get with it out there."

Then, at 6:31 P.M., these words came over the radios of the technicians on the ground: "Fire. I smell fire." A television camera trained on the small window in the command module recorded a bright flash. There was a scream over the radios. One of the astronauts shouted: "We've got a fire in the cockpit."

After 14 seconds, the hatch on the overheated capsule buckled, allowing black smoke to escape. Still, it took six minutes for technicians to pry open the hatch to *Apollo 1*. Inside, they found the charred remains of the three astronauts. They had died within seconds.

Grissom, White, and Chaffee were the first astronauts to die while participating in the American space program. Since 1961, America had sent 16 manned missions into space. Twenty-six astronauts had flown in the Mercury and Gemini spacecrafts. There had been mishaps on some of those missions; nevertheless, the astronauts in space or the engineers on the ground had always come up with ways to solve the problems.

The objective of the Mercury program, which featured a single astronaut aboard a space capsule, was simply to achieve **orbital space flight**. The two-man Gemini program was intended to develop and test the skills that would be needed on a flight to the moon. Space walks by the astronauts and long-duration flight were part of the Gemini mission plan. So was the technique of **rendezvous**–a process by which two spacecraft would link in orbit. All those objectives had been completed.

These photos show the severe damage to the interior (inset) and exterior of the Apollo 1 *capsule. The fire set back America's space program for more than a year, as NASA tried to determine what had gone wrong and correct the problems.*

The missions of the Mercury and Gemini programs had been accomplished. Project Apollo was to be the next, and final, step to the moon. But now, on Pad 34, the Apollo program lay in ruins.

What went wrong? Hundreds of NASA investigators were assigned the job of figuring out what caused the fire, and how it managed to burn so quickly and completely. They learned the fire had started near a bundle of wires in front of Grissom's seat. A spark from one of the wires had apparently ignited flammable materials in the command module. The fire was fed by the oxygen-rich atmosphere inside the capsule.

Over the next 21 months, many changes were made to the Apollo capsule. Wiring was rerouted and better insulated. The atmosphere in the capsule was changed to a less dangerous mixture of oxygen and nitrogen. The hatch was redesigned, so that it could be opened quickly by the astronauts.

The investigation and the changes made to the Apollo spacecraft slowed down the American space program. Meanwhile, the Soviet Union's space program was proceeding swiftly. Soviet **cosmonauts** were carrying out missions in Earth orbit.

But while work on the capsule was being completed, NASA was testing a powerful new rocket called the Saturn V. The first unmanned Saturn V was successfully launched on November 9, 1967. NASA then launched four more unmanned Apollo flights to test the rocket.

On October 11, 1968, three astronauts climbed aboard the *Apollo 7* spacecraft. Their flight would have modest goals—astronauts Walter Cunningham, Donn Eisele, and Wally Schirra were to take the command module on a shakedown cruise in Earth orbit and perform a simple rendezvous with one of the Saturn booster stages.

During the Mercury and Gemini space programs, the astronauts were launched into space on rockets that had been designed as missiles. To break out of Earth's gravitational pull, a more powerful rocket was needed: the Saturn V.

"We just slid right up the pipe and onto the target," reported Eisele. "It was a great feeling."

The mission was a success, and the American space program again aimed for the moon.

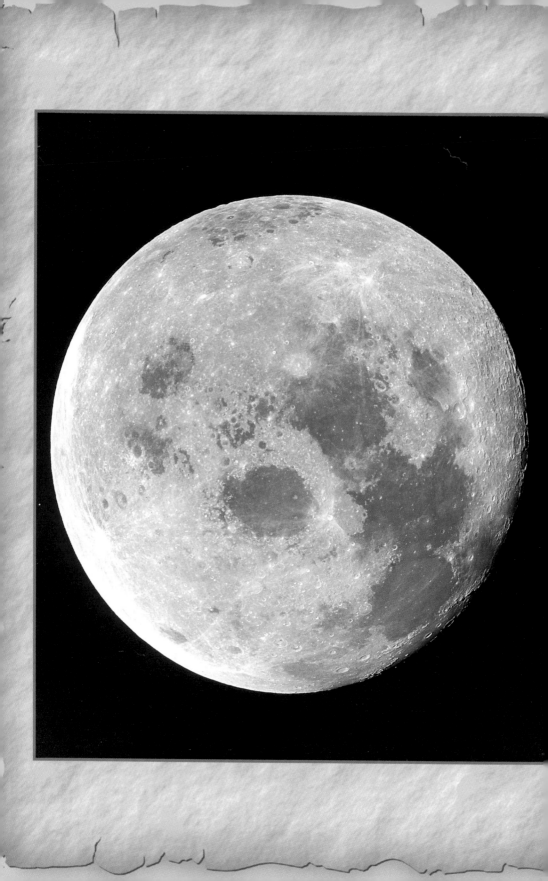

For thousands of years, humans have looked up at the night sky, wondered about the moon, and dreamed of one day visiting Earth's bright satellite.

A Nation Commits Itself

2

he path to the moon did not start on a launching pad in sunny Florida. Rather, the idea of shooting a manned rocket to the moon had its beginnings in a tiny windswept village in the Kaluga province of Russia. The local schoolteacher there was a lonely, shy, and deaf man named Konstantin Eduardovich Tsiolkovsky. He was born in September 1857.

Tsiolkovsky did not come up with the idea of the self-propelled rocket. For centuries, rockets had been fired in battle. The first recorded use of them as a weapon was in the 13th century, when the Chinese ignited small rockets to repel an invading Mongol army. Nor was Tsiolkovsky the first to envision the idea of space flight–French author Jules Verne's 1865 novel *From the Earth to the Moon* suggested that a manned space capsule could be propelled to the moon by being shot out of a giant cannon. Other science fiction writers came up with similar schemes.

But working in Kaluga on his own, Tsiolkovsky essentially mapped out theories of space flight. He published papers in the early 1900s suggesting that a rocket propelled by liquid fuels could break the gravitational pull of the Earth and reach the moon. He even suggested that the fuels could be composed of liquid hydrogen and oxygen. (Both of these ***propellants*** would become major components of the fuels used in rockets.) Tsiolkovsky also recommended the use of multi-stage rockets that would conserve fuel by shedding heavy parts when they were no longer of use.

Because of his writings about space flight, Konstantin Eduardovich Tsiolkovsky would become known as the Father of Astronautics.

Tsiolkovsky died in 1935, two decades before any rocket was shot into space. Nevertheless, the following words were etched in his grave marker: "Man will not stay on Earth forever, but in pursuit of light and space will emerge timidly from the bounds of atmosphere and then advance until he has conquered the whole of circumsolar space."

While Tsiolkovsky was living and working in rural Russia, two young men in other parts of the world were also dreaming about space flight. One was an American, Robert Hutchings Goddard; the other a German, Wernher von Braun.

Goddard was born in 1882 in Worcester, Massachusetts. He grew up reading science-fiction books about space flight and was convinced it could be accomplished. By the early 1920s, Goddard was a professor of physics at Clark University in Massachusetts. He was also experimenting with rockets and propellants.

In March 1926, on his aunt Effie's farm near Auburn, Massachusetts, Goddard launched the world's first liquid-propellant rocket. The small rocket reached a height of only 41 feet and a distance of just 184 feet. It is hard to imagine that even Goddard could envision that tiny accomplishment as the first step toward a mission to the moon. Nevertheless, American rocketry had been born.

Ten years later, Goddard would be heading teams that would build rockets capable of reaching altitudes of some 9,000 feet and speeds approaching supersonic levels. Those were very impressive advancements, but they did not come near the speeds and altitudes needed to achieve space flight.

Meanwhile, in Germany Wernher von Braun and a group of space flight enthusiasts were testing their own rockets. Their work came to the attention of Nazi dictator Adolf Hitler, who could see the military uses of a German rocketry program. Soon, von Braun and the other rocket scientists found themselves working for the German army.

During World War II, von Braun and his fellow rocket scientists developed two terrible weapons— the V-1 and V-2 rockets. These were armed with explosives and unleashed against London. But although the rockets caused massive destruction in the city, they did not alter the course of the war. By the time the Nazis were able to employ the V-1 and V-2 rockets, the war was nearly over.

As the Allied forces advanced into Germany, von Braun called Germany's top rocket scientists together. They agreed to let themselves be captured by the Americans, because they believed America

German rocket scientist Wernher von Braun became an important part of America's space program. He had been captured by U.S. forces during World War II. When asked what it would take for men to reach the moon, he answered, "The will to do it."

would one day launch a space program. Von Braun and many of his assistants surrendered to American forces. They soon found themselves working in laboratories and testing grounds in the United States as the American rocket program was born.

For America, the next great leap forward in the space race occurred early in the morning of October 14, 1947. Air force captain Chuck Yeager, at the controls of the X-1 rocket plane, became the first man to break the sound barrier, flying at a speed greater than 750 miles an hour. It had been believed that no plane could fly faster than the sound barrier (known as **Mach 1**), and that the body of the plane would break up due to the tremendous vibrations that shook the aircraft at such high velocities. But Yeager soared right through Mach 1, and soon other pilots would break speed records as well.

Of course, American engineers were nowhere near designing planes, or rockets, that could achieve the speeds needed to break the Earth's gravity and enter space. The X-1 had traveled more than 750 miles an hour. To break the Earth's gravitational pull, a rocket would have to achieve a speed greater than 24,000 miles an hour.

America concentrated on building airplanes that would travel faster and higher, such as the X-15. Meanwhile, though, a space program was growing in the Soviet Union. The Soviets were developing liquid-fuel rocket boosters that could propel *payloads* into Earth orbit. They achieved this goal on October 4, 1957, by successfully launching a small satellite, *Sputnik I*, into orbit. *Sputnik* was about the size of a basketball, and it did nothing more than transmit a radio signal back to Earth. However, its presence in the sky upset Americans, who thought they would be first in space.

The decades after World War II were a time of great tension between the two world powers. Both countries had nuclear weapons, and it was feared that, eventually, spacecraft could be used to deliver atomic bombs on an enemy. Government leaders urged President Dwight D. Eisenhower to begin a national space program that could compete with the Soviet Union. Eisenhower responded by establishing the National Aeronautics and Space Administration (NASA) in 1958. Officials at the new space agency soon developed America's first manned space program. It would be called Project Mercury. Seven military test pilots were selected as members of the astronaut corps, and they quickly immersed themselves in the training for their missions.

But the Soviet Union would beat America again!

On April 12, 1961, a Soviet space capsule was launched into orbit carrying

More than 100 of the country's best test pilots were considered for the Mercury space program. Seven astronauts were selected: Walter M. Schirra Jr., Donald K. "Deke" Slayton, John H. Glenn Jr., M. Scott Carpenter, Alan B. Shepard Jr., Virgil I. "Gus" Grissom, and L. Gordon Cooper.

cosmonaut Yury Gagarin. His spacecraft, *Vostok 1*, made a single orbit of the Earth. The Russians had not yet figured out a way for *Vostok 1* to land safely, so after the capsule reentered the atmosphere Gagarin was forced to bail out and parachute to the ground. Still, the Soviet Union had shot a man into Earth orbit!

Three weeks later, on May 5, 1961, Alan B. Shepard Jr. became the first American to fly in space. The flight of his spacecraft, *Freedom 7*, was **suborbital**, meaning it simply flew in a small arc after liftoff before re-entering Earth's atmosphere. Nevertheless, the success of *Freedom 7* was a giant step forward for the American space program. And Shepard had stayed in the spacecraft all the way to **splashdown**—there was no need for him to jump out of the capsule, as Gagarin had done.

The man most impressed with Shepard's flight was President John F. Kennedy. On May 25, 1961, just three weeks after the flight of *Freedom 7*, President Kennedy addressed a joint session of Congress in Washington. He said, "I believe that this nation should commit itself to achieving the goal, before this decade is out, of landing a man on the moon and returning him safely to Earth. No single space

American astronaut Alan Shepard is cheered by the crew of the USS Lake Champlain, *the rescue ship that plucked him out of the ocean after his space flight. Behind him is the* Freedom 7 *capsule.*

project in this period will be more impressive to mankind or more important for the long-range exploration of space. And none will be so difficult or expensive to accomplish."

A NASA artist's conception of a rendezvous between the lunar module (top) and the command service module. Docking two ships in space was tricky, but it was the most effective way to reach the moon and return.

Lunar Orbit Rendezvous 3

For most Americans, the notion of what a moon mission was all about emerged from the science-fiction movies of the 1950s. These films would show a tall rocket ship that would blast off from a launch pad on Earth. Its rockets would enable it to soar the 240,000 miles to the moon. Fire would stream from the shiny ship's finned tail as it raced through space. Once above the moon, the ship's pilot would maneuver the spacecraft so that the tail would point toward the lunar surface. Then, with landing rockets blazing, he would deftly ease the huge craft to a soft landing. The return trip to Earth would be handled much the same way: the craft would blast off from the moon, soar back to Earth, and then make a soft landing on the same launch pad where the mission started.

In reality, things were much more complicated than that. For starters, space mission planners had adopted Konstantin Tsiolkovsky's idea about shedding rocket stages as they were drained of their fuel. This was a concept that had been employed with the first satellite launches in the 1950s and used throughout the Mercury and Gemini programs. The rocket and capsule weighed approximately six million pounds, and there simply was not enough fuel to send it all the way to the moon and back.

For the Apollo missions, the astronauts' spacecraft would sit atop the enormous Saturn V rocket. The 363-foot-tall Saturn V had three stages. The first was needed to lift the rocket and astronauts off the launch pad. Providing 7.5 million pounds of thrust—equivalent to 180 million horsepower—it sent the rocket 36 miles into the air. When the first stage burned out after two and a half minutes and dropped away, the second stage would fire. This sent the rocket into earth orbit at a speed of more than 17,000 miles per hour. Finally, the third stage

The enormous Saturn V rocket sits on a launching pad.
The three-stage rocket weighed six million tons.

was fired to help the spaceship reach a velocity of
nearly 25,000 miles per hour. This allowed the
astronauts to break away from Earth's gravitational
pull and head toward the moon.

The third stage also carried the **lunar module** (LM). This odd-looking craft would carry the astronauts to the surface of the moon. Above that was the **service module** (SM), which contained the air supply for the astronauts as well as fuel for the voyage and a rocket engine that would help the craft make navigational adjustments to get to the moon and back. This was attached to the command module (CM), which carried the astronauts. This combined craft was known as the command service module (CSM).

Once the Saturn V rocket had sent the astronauts out of Earth orbit, they would reach the moon by a method called **lunar orbit rendezvous**. The idea had been developed by an engineer named Dr. John C. Houbolt. Initially there was some resistance to the idea, because it involved using two spacecrafts and seemed very complicated. However, NASA soon became convinced that lunar orbit rendezvous was the most efficient way to get to the moon.

First, the command service module would break away from the booster stage carrying the lunar module. Then, the astronauts inside the CSM would turn the craft around so that it faced the drifting booster stage containing the LM. The pilot of the command service module then had to perform a complicated maneuver that would lock the nose of the spacecraft into a port on the lunar module. Next, the CSM would pull back, withdrawing the LM from the third stage. After that was accomplished, the astronauts could begin the three-day trip to the moon.

Once the astronauts achieved lunar orbit, two of the three Apollo astronauts would enter the LM. The command service module pilot would remain in the CSM, orbiting the moon. The lunar module would then break away from the command module and begin the descent to the surface of the moon. The LM was powered by a small rocket engine. It had four legs that extended below the craft, giving the lunar module the appearance of a squatting bug. The two astronauts would fly to the designated landing spot and find a safe place to set their craft down.

When the astronauts finished their work on the moon, the top portion of the lunar module would blast off from the bottom section.

> "LOR [lunar orbit rendezvous] offered a chain reaction of simplifications," NASA engineer Dr. John C. Houbolt, who came up with the concept, has explained. "Development, testing, manufacturing, launch, and flight operations all would be simplified."

The lunar module projects from the third stage of a Saturn V rocket. Before the stage could drift away, the astronauts had to turn the command service module around, dock with the LM, and extract it from the expended stage.

Once back in lunar orbit, the LM would link up with the CSM. The astronauts would return to the command module, bringing along the rocks and soil samples they had collected on the surface. Once the hatch between the two ships was sealed, they could discard the LM. Next, the astronauts would fire the CSM's rocket engine to break the gravitational pull of the moon and head for Earth. Although the engine burn would last only a few minutes, it would be enough to propel the astronauts back home.

In Earth orbit, the command module would separate from the service module. The capsule would then descend through the atmosphere, deploy parachutes to slow its fall, and splash down in the Pacific Ocean, where it would be recovered by U.S. Navy ships. All that would be left of the enormous rocket that had blasted off a week before would be the 11-foot-tall gumdrop-shaped command module bobbing in the choppy waters of the Pacific.

The theory was sound, but NASA was not taking chances. The Apollo program would move slowly, carefully testing at each stage and learning from problems and mistakes. After all, the Saturn V rocket and Apollo spacecraft had more than six million moving parts. All of them had to work together perfectly for the mission to be a success.

After the setback with *Apollo 1*, NASA successfully tested the Saturn V rocket. The command service module underwent an overhaul to make it safer. When these were ready, it was time to begin manned space flights again.

First, Cunningham, Eisele, and Schirra took *Apollo 7* aloft and found the spacecraft's performance to be perfect. They also practiced the complicated docking maneuver with the discarded third booster stage. This would simulate docking with the LM, which would be necessary in later missions.

Apollo 8 was next. Astronauts Frank Borman, James Lovell, and William Anders took their spacecraft all the way to the moon, 240,000 miles away. They flew around it, sending back closeup photos on Christmas Eve 1968, then returned. This proved that both the spacecraft and its human crew could survive the week-long journey.

Apollo 9's mission was conducted in Earth orbit. Astronauts James McDivitt, David Scott, and Russell Schweickart carried out an actual docking with the lunar module. All seemed nearly ready, but NASA decided it needed one more trial run before attempting an actual moon landing. *Apollo 10* was blasted into lunar orbit, where astronauts Thomas Stafford, Eugene Cernan, and John Young put the lunar orbit rendezvous plan to the test. The LM, with Stafford and Cernan aboard, was cut loose from the command service module. It descended to within nine miles of the lunar surface, then returned to the CSM. Lunar orbit rendezvous worked.

On this mission, the astronauts also showed they could deal with unexpected situations. During the LM's flight, a switch was thrown into the wrong position, causing the craft to shake violently back and forth. But Stafford quickly figured out the problem, and he wrestled the lander back onto a steady course.

With that problem out of the way, the astronauts could take a moment to gaze upon the barren moonscape that sprawled out below them. Cernan could hardly contain his enthusiasm. "We're right there! We're right over it," he told Mission Control back on Earth. "I'm telling you, we are low, we're close, babe."

> **To prepare for the first moon landing, NASA spent $40 billion on the Mercury, Gemini, and Apollo programs. Some 20,000 private companies received contracts to build components of the Apollo spacecraft. More than 400,000 people working for those companies contributed in some way to the development and construction of the Apollo spacecraft.**

This view of Apollo 11 *lifting off from Cape Canaveral on July 16, 1969, was taken by a camera on the gantry.*

"The Eagle Has Landed" 4

Neil Armstrong was born on an Ohio farm, but even as a young boy Armstrong knew his future would not be working in his father's grain fields. Neil wanted to fly—an ambition he fulfilled by the age of 16, when he earned his pilot's license. He went on to enlist in the navy and became a fighter pilot. Armstrong flew 78 missions during the Korean War. After the war, Armstrong attended Purdue University, where he completed a degree in aeronautical engineering. Soon, he found himself working as a civilian test pilot for NASA. In 1962, he joined NASA's astronaut corps and flew Gemini missions.

Edwin "Buzz" Aldrin Jr. attended the U.S. Military Academy at West Point, then enlisted in the air force. Like Armstrong, he flew fighter jets during the Korean War. But Aldrin's interest in flight went far beyond just working the controls of a speeding jet plane. After the war, he obtained a doctoral degree in engineering. By the time he joined the astronaut corps in 1963, Aldrin knew more about how to rendezvous two spacecraft than anyone else at NASA. In fact, his nickname was "Dr. Rendezvous." He made his first space flight aboard *Gemini 12* in 1966.

Michael Collins, another West Point graduate, had an army background. His father, General James L. Collins, was a respected military leader who had helped track down the Mexican outlaw Pancho Villa in 1916. Collins, however, decided to become a fighter pilot, and in 1963 he was accepted into NASA's astronaut corps.

NASA selected Armstrong, Aldrin, and Collins as the crew of *Apollo 11*, the first mission that would actually land on the moon. Armstrong was named commander of the mission. He and Aldrin would take the LM, which they had named *Eagle*, down to the moon's surface. Collins, meanwhile, would orbit the moon in the command module *Columbia*.

Neil Armstrong (left) was the commander of Apollo 11.
*Next to him are Michael Collins (center), who would fly
the CSM in orbit around the moon, and Edwin "Buzz"
Aldrin, who would join Armstrong on the lunar surface.*

The launch was scheduled for July 16, 1969, with
the moon landing expected on July 20. If the mission launched on schedule, *Apollo 11* would fulfill
President Kennedy's goal of landing on the moon
before the end of the 1960s. "After a decade of plan-
ning and hard work, we're willing and ready to

achieve our national goal," Armstrong said a few days before liftoff.

Apollo 11 lifted off for the moon at 9:32 A.M. It seemed as though the entire Florida peninsula shuddered as the huge Saturn V rocket ignited.

"We're thrown left and right against our straps in spasmodic little jerks," Collins told Mission Control, which was based in Houston, Texas. "It is steering like crazy . . . I just hope it knows where it's going."

Eleven minutes later, *Apollo 11* achieved Earth orbit. After one and a half orbits, the astronauts fired the third-stage booster and the spacecraft headed for the moon. Once the burn was completed, their next job was to separate *Columbia* from the third stage,

A special patch was created for the Apollo 11 *astronauts to wear on their space suits.*

and link up with *Eagle.* This would be Collins's job.

"This was a critical maneuver in the flight plan," Buzz Aldrin later said. "If the separation and docking did not work, we would return to Earth. There was also the possibility of an in-space collision and the subsequent decompression of the cabin. Critical as the maneuver is, I felt no apprehension about it, and if there was the slightest inkling of concern it disappeared quickly as the entire separation and docking proceeded perfectly to completion."

Columbia and *Eagle* were now linked and headed for the moon. There was little for the astronauts to do now but relax. The moon landing was still three days off. As time passed, they saw the moon growing larger and larger. "The moon I have known all my life–that two-dimensional small yellow disk in the sky–has gone away and been replaced by the most awesome sphere I have ever seen," Collins told Mission Control. "To begin with, it's huge–completely filling our window. Second, it is three-dimensional. The belly of it bulges out toward us in such a pronounced fashion that I almost feel I can reach out and touch it."

When *Columbia* achieved lunar orbit, Armstrong and Aldrin made their way into the LM *Eagle.* They prepared to separate the LM from the CSM. Armstrong's final words to Collins were, "See you later." Then Aldrin fired the thrusters on *Eagle*, separating the lunar lander from *Columbia*. The descent toward the moon had started. Armstrong took the controls of the lunar module.

Eagle was aiming for the Sea of Tranquility. This is a large flat plain on the surface of the moon. The moon has no atmosphere and no water, so there can be no actual "seas." But centuries ago, astronomers studying the moon through primitive telescopes saw dark regions on its surface. Believing these were large bodies of water, they named them seas.

"How does it look?" Mission Control asked the LM. "*Eagle* has wings," Armstrong responded.

At just over a mile above the lunar surface Aldrin noticed a yellow caution light flash on *Eagle*'s control panel. Moments later, a second yellow light appeared. The caution lights meant that the ship's onboard computer had become overloaded–it was being asked to do too much–and that it would delay carrying out its functions until its memory was able to handle the influx of data.

Today, personal computers often slow down if they are given too many commands at once. Of

Michael Collins took this photo of the separated LM,
with the Earth rising over the moon in the background.

course, when a personal computer slows down, it
hardly presents a life-or-death situation. On board
Eagle, however, a computer slowdown meant that

control of the spacecraft would be hampered because the computer would not carry out the pilot's orders fast enough.

"Our hearts shot up into our throats while we waited to see what would happen," recalled Aldrin.

That's when Steve Bales stepped in. Back at Mission Control in Houston, Bales was responsible for the computer aboard the LM. Bales told Armstrong and Aldrin to ignore the warning lights and carry on with the descent. He had determined that there was nothing wrong with the computer. Later, Bales would admit that he gave the astronauts the go-ahead to continue the descent "on instinct."

Eagle was now just a few feet from the surface of the moon. "Thirty seconds," Mission Control said, meaning *Eagle* had 30 seconds of fuel remaining for the descent.

Aldrin watched the control panel. A rod extending from the bottom of the LM touched the lunar surface. A light appeared on the panel, showing the rod had made contact. "Contact light," he said.

"We copy you down, *Eagle*," said Mission Control. There was a pause of a few seconds.

"Houston, Tranquility Base here," Armstrong said. "The *Eagle* has landed!"

Mission Control erupted into cheers. All across America and the world, millions of people watching on TV or listening to their radios had heard those words.

"Roger, Tranquility. We copy you on the ground," Mission Control answered. "You've got a bunch of guys about to turn blue. We're breathing again. Thanks a lot."

A close-up view of an Apollo 11 *astronaut's bootprint on the moon's powdery surface.*

"That's One Small Step . . ."

5

$\mathcal{A}$rmstrong and Aldrin took a moment to look through the small windows of the LM. The dust on the surface of the moon had been stirred up by the landing. When it settled, the two astronauts could see the lunar terrain spread out before them. "We saw a surface pockmarked with **craters** up to 15, 20, 30 feet, and many smaller craters down to a diameter of one foot and, of course, the surface was very fine-grained. There were a surprising number of rocks of all sizes," Armstrong said.

Craters are formed when a meteorite hits the moon or a planet. When a meteorite strikes a rocky moon or planet, the impact on the surface leaves a depression in the surface. The sides of a crater resemble a wall.

> Most craters on the moon are believed to be millions of years old. They were formed at a time when meteorites were a lot more numerous than they are now.

The Earth regularly travels through meteor showers. However, most meteorites burn up when they enter the atmosphere. Friction causes them to disintegrate well before they strike the ground. There have been exceptions, however. Near Winslow, Arizona, the Barringer Meteorite Crater measures nearly a mile across.

There are some craters on the moon many times that size. Armstrong and Aldrin did not see any of them close up, though. They had, after all, chosen to land in the Sea of Tranquility.

Armstrong continued his description of the lunar surface. "It's pretty much without color," he said. "It's gray and it's a very white chalky gray."

It was now about 5 P.M. in Houston. Armstrong and Aldrin were supposed to rest after the excitement of the descent and landing, but neither man was tired. They told Mission Control in Houston that they intended to dress in their moon-walking suits and leave the LM as soon as possible. Mission Control gave them the go-ahead. The walk on the moon—known as Extravehicular Activity (EVA)—was planned to begin in three hours.

There were delays, though. The two astronauts thought it would take two hours to put on their bulky moon suits. Instead it took closer to four hours. Also, the astronauts were having difficulty with the controls that bled the air out of the *Eagle* cabin. With no atmosphere on the moon, they had to equalize the air pressure inside the LM with the vacuum of space outside the craft before they could open the hatch. Finally, that task was completed.

It was now nearly 10 P.M. in Houston. Armstrong and Aldrin had been on the moon for almost five hours. The hatch on the roof of the LM opened. Armstrong climbed out and paused for a moment on a ledge above *Eagle*'s ladder. Before heading down the nine rungs of the ladder, he pulled open a hatch on the side of *Eagle* that contained some equipment the astronauts would use while they were on the surface of the moon. It also contained a television camera.

The camera activated when Armstrong opened the equipment hatch. Suddenly, everyone watching

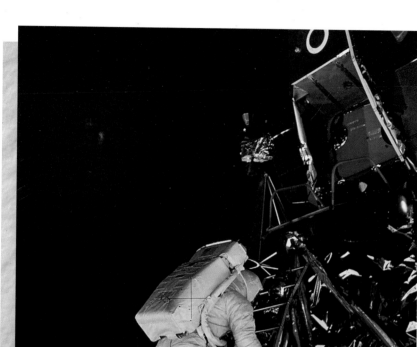

Neil Armstrong, standing on the moon's surface, snapped this photo of Buzz Aldrin stepping off the LM's ladder.

their TV sets on Earth saw an image of Neil Armstrong perched on the top of *Eagle's* ladder. The image was in black and white; nevertheless, it meant that millions of people on Earth could now share in the experience of watching the first men walk on the moon, 240,000 miles away.

"Okay, Neil, we can see you coming down the ladder," Mission Control reported.

Armstrong stepped carefully down the ladder. Next, he hopped onto the base of the LM's landing gear. "I'm going to step off the LM now," he said.

At 9:56 P.M. Houston time, Neil Armstrong placed a foot into the ashy white dust of the moon. He said, "That's one small step for man, one giant leap for mankind."

As Armstrong took several more steps away from the LM, he found it surprisingly easy to walk on the moon. Even though he was dressed in a heavy, cumbersome space suit, the moon has just one-sixth the gravity of Earth. Therefore, the suit felt much lighter than it would on Earth.

"I am standing directly in the shadow now looking up at Buzz in the window and I can see everything quite clearly," Armstrong told the enthralled viewers. "The light is sufficiently bright. Everything is quite clearly visible."

It was now Aldrin's turn to exit the LM. He, too, made his way out of the LM hatch and down the ladder. When he stepped onto the surface he looked around at the stark lunar landscape. "Magnificent desolation," he commented.

The two astronauts soon got down to work. They collected rock and dirt samples. Next, they unveiled a plaque that was attached to one of the legs of the LM. It would remain on the moon after the two men blasted off. The plaque showed the two **hemispheres** of the Earth, the signatures of Armstrong, Aldrin, and Collins, and the signature of President Richard M. Nixon. It also said, "Here men from the planet Earth first set foot upon the Moon, July 1969, A.D. We came in peace for all mankind."

Their next job was to raise an American flag. Aldrin and Armstrong had some difficulty getting the flagpole to stand in the soft lunar dust. Finally the pole stood on its own. Because there is no wind on the moon, the pole contained an arm that would hold the flag erect. Armstrong, who had a camera mounted on the front of his moon suit, took a picture of Aldrin saluting the flag.

Next, Mission Control told the astronauts that President Nixon was ready to speak with them. Nixon's call to the astronauts was patched through from the White House in Washington, D.C.

"This certainly has to be the most historic tele-

This plaque was mounted on the Apollo 11 *lunar module to mark the spot where men first set foot on the moon. It was signed by the three* Apollo 11 *astronauts, as well as President Richard M. Nixon.*

phone call ever made," the president said. "I just can't tell you how proud we all are of what you have done. For every American, this has to be the proudest day of our lives. And for people all over the world, I am sure that they, too, join with Americans in recognizing what an immense feat this is. Because of what you have done, the heavens have become a part of man's world."

"Thank you, Mr. President," Armstrong said. "It's a great honor and privilege for us to be here representing not only the United States but men of peace of all nations—and with interest and curiosity, and men with a vision for the future. It's an honor to be able to participate here today."

Buzz Aldrin sets up the scientific instruments that the astronauts would leave behind on the moon. In the background are the lunar module and the American flag.

After the phone call, the astronauts still had work to do. They set up the Early Apollo Scientific Experiments Package—an array of scientific instruments. One was a **seismograph** that would measure "moon quakes." Another was a laser reflector that would enable astronomers to measure the exact distance between the Earth and the moon. They also set up a solar wind detector, which would trap and analyze gases emitted by the sun.

Soon, Mission Control ordered the astronauts back into the LM. They still had plenty of air left in their suits, but Mission Control was concerned that Armstrong and Aldrin might become tired. They had been going nonstop since waking up in lunar orbit 18 hours earlier. It took three hours for them to get back into the LM, pressurize the cabin, and remove their moon suits.

Armstrong and Aldrin were supposed to sleep for seven hours on the moon, but *Eagle* contained no bunks. Aldrin tried to sleep on the floor of the tiny lunar module while Armstrong climbed into a hammock, but the two men were unable to get any rest. Both men reported that the LM was too chilly for them to sleep.

Finally, though, the moment came for *Eagle* to take off from the moon. The rocket fired and the tiny LM soared back into lunar orbit, where it met up with *Columbia.* After a successful rendezvous, Armstrong, Aldrin, and Collins were on their way home.

A perfect liftoff for Apollo 13 *on April 11, 1970. Things did not stay perfect, however; an explosion in space forced the astronauts to return without landing on the moon.*

For All Mankind 6

ix more moon missions would follow *Apollo 11*. Ten more men would walk on the lunar surface over the next three years. They would conduct dozens of scientific experiments and bring home several hundred pounds of moon rocks. Alan Shepard, the first American in space, would practice his golf swing on the moon during the *Apollo 14* mission. Three crews would drive the **moon rover**, an electric cart that would enable the astronauts to travel miles from their landing sites. During one mission, the astronauts retrieved a moon rock believed to be 4.5 billion years old.

Not all the missions were successful. On April 11, 1970, *Apollo 13* lifted off for the moon. Two days later, when the spacecraft was 200,000 miles away from Earth, an oxygen tank in the service module exploded. The power quickly drained out of the command module. To survive, astronauts James A. Lovell Jr., John L. "Jack" Swigart Jr., and Fred W. Haise Jr. had to rely on the LM's life-support systems as they returned to Earth.

After several days in space, *Apollo 13* approached Earth orbit. The service module was cut loose. As it drifted away, the astronauts saw through their window that one side had been blown away by the explosion. The crew then guided the command module through the atmosphere. The spacecraft splashed down in the Pacific Ocean, and the astronauts were recovered safely.

An investigation showed that some wires inside an oxygen tank had overheated during a long test on the launch pad, causing insulation to burn away. When the wires overheated again, this time in space, a spark caused the tank to explode.

Despite this setback, NASA fixed the problem and the Apollo program continued. Finally, on December 11, 1972, the LM of *Apollo 17* touched

Apollo 15 *astronaut James B. Irwin works at the moon rover. During the Apollo program, 12 Americans set foot on the moon's surface.*

down on the moon. Astronauts Eugene Cernan and Harrison H. Schmidt spent 75 hours on the moon. They drove their moon rover more than 20 miles, exploring the lunar surface. They made three visits to the surface of the moon, including one that lasted more than seven hours. Cernan and Schmidt also left behind a plaque on the moon. It reads: "Here man completed his first explorations of the moon. May the spirit of peace in which we came be reflected in the lives of all mankind."

Chronology

1857 Konstantin Eduardovich Tsiolkovsky is born in Kaluga, Russia. The obscure schoolteacher will become known as the Father of Astronautics for establishing theories of spaceflight that will prove to be true.

1865 French author Jules Verne's novel *From the Earth to the Moon* proposes that men could travel to the moon inside a space capsule.

1926 On his aunt's farm near Auburn, Massachusetts, Robert Hutchings Goddard launches the world's first liquid-propellant rocket on March 16. The small rocket reaches a modest height of 41 feet and flies a distance of 184 feet.

1945 As World War II draws to a close, Wernher von Braun and other German rocket scientists surrender to American troops. The German scientists are taken to the United States, where they are put to work in laboratories and missile testing grounds.

1947 On October 14 Chuck Yeager, a captain in the air force, breaks the sound barrier while flying the X-1, an experimental jet. By flying the X-1 beyond Mach 1 (approximately 750 miles an hour) Yeager proves that aircraft can withstand the punishment of high-speed flight.

1957 The unmanned satellite *Sputnik I* is launched by the Soviet Union on October 4.

1961 Soviet Cosmonaut Yury Gagarin becomes the first man in space when he completes one orbit of the Earth in the *Vostok 1* spacecraft on April 12; Alan B. Shepard Jr.

becomes the first American to fly in space in *Freedom 7* on May 5; during a speech on May 25, President John F. Kennedy commits the nation to land a man on the moon before the end of the decade.

1967   Astronauts Virgil I. "Gus" Grissom, Edward H. White, and Roger B. Chaffee are killed when fire sweeps through the *Apollo 1* capsule during a test on a Cape Canaveral launch pad on January 27.

1968   *Apollo 7* astronauts Walter Cunningham, Donn Eisele, and Wally Schirra complete the first successful manned Apollo flight in October.

1969   *Apollo 11* crew members Neil Armstrong and Edwin "Buzz" Aldrin Jr. step out of the lunar module *Eagle* onto the surface of the moon on July 20, then return to Earth in the command module *Columbia*, piloted by Michael Collins.

1970   After an oxygen tank explodes in the service module of *Apollo 13*, disabling the spacecraft, astronauts James A. Lovell Jr., Fred W. Haise Jr., and John L. "Jack" Swigart Jr. use the lunar module as a lifeboat, propelling the crippled command module around the moon and back.

1972   *Apollo 17* splashes down in the Pacific Ocean on December 19, ending America's manned exploration of the moon.

Glossary

astronaut–a space traveler.

command module–the 11-foot-high capsule in which the Apollo astronauts left and reentered the Earth's atmosphere.

cosmonaut–the Russian name for a space traveler.

crater–a cup-shaped depression resulting from the impact of a meteorite.

hemisphere–half of a round object, such as a planet.

lunar module–a special lightweight craft used to transport astronauts to and from the moon's surface.

lunar orbit rendezvous–the method NASA used to land on the moon. When the Apollo spacecraft reached the moon, the lunar module would separate from the command service module. The LM would land on the surface while the CSM orbited the moon. When it was time to return, the LM would take off from the moon and rendezvous with the CSM. After the astronauts exited the LM, the craft would be jettisoned and all would return to earth in the CSM.

Mach 1–a term that designates the speed of sound. The speed needed to break the "sound barrier" varies depending on the temperature and atmospheric level of travel, but it is approximately 750 miles per hour.

moon rover–a four-wheeled vehicle used by Apollo astronauts to explore greater areas of the moon's surface.

orbital space flight–a flight into space in which the craft orbits, or circles, a large body such as the Earth or the moon.

payload–the satellite or spacecraft that is carried into space by a rocket.

propellant–fuel used by a rocket engine.

rendezvous–a connection or docking between two spacecrafts.

rocket–a jet engine that carries its own fuel and oxygen for combustion, and is used to propel a vehicle, such as a space capsule, through the air.

seismograph–an instrument that records the distance and intensity of tremors on the Earth or moon.

service module–a cylinder connected to the command module that contained oxygen and a rocket booster that would be used by the astronauts during the trip to and from the moon. The service module was discarded before the command module reentered Earth's atmosphere.

splashdown–the landing of a manned spacecraft in the ocean.

suborbital space flight–a flight into space that does not orbit the Earth, but travels out of and back into the atmosphere in an arc.

Further Reading

Bredeson, Carmen. *Neil Armstrong: A Space Biography*. Springfield, N.J.: Enslow Publishers, 1998.

Chaikin, Andrew. *A Man on the Moon*. New York: Viking, 1994.

Furniss, Tim. *One Giant Leap: The Extraordinary Story of the Moon Landings*. London: Carlton Books, 1999.

Hasday, Judy. *The Apollo 13 Mission*. Philadelphia: Chelsea House Publishers, 2001.

Hurt, Harry. *For All Mankind*. New York: The Atlantic Monthly Press, 1988.

Kennedy, Gregory P. *Apollo to the Moon*. New York: Chelsea House Publishers, 1992.

Mason, Robert Grant. *Life in Space*. Boston, Mass.: Little, Brown and Company, 1983.

Wilford, John Noble. *We Reach the Moon*. New York: Bantam Books, 1969.

Wolfe, Tom. *The Right Stuff*. New York: Farrar, Straus and Giroux, 1979.

Picture Credits

HAL MARCOVITZ is a reporter for the *Allentown (Pa.) Morning Call.* His work for Chelsea House includes biographies of explorers Marco Polo, Francisco Vazquez de Coronado, and the Lewis and Clark Expedition guide Sacajawea. He has also written a biography of the actor Robin Williams, as well as a history of terrorism in America. He is the author of the satirical novel *Painting the White House.*